SCIENTIFIC AMERICAN | EDUCATIONAL PUBLISHING

SCIENTIFIC AMERICAN INVESTIGATES GEMS

EMERALDS

BY MADELEINE FORTESCUE

Published in 2027 by The Rosen Publishing Group
in association with Scientific American Educational Publishing
2544 Clinton Street, Buffalo NY 14224

Portions of this work were originally authored by Eric Ethan and published as *Emeralds*. All new material in this edition is authored by Madeleine Fortescue.

Cataloging-in-Publication Data
Names: Fortescue, Madeleine.
Title: Emeralds / Madeleine Fortescue.
Description: Buffalo, New York : Scientific American Educational Publishing, an imprint of Rosen Publishing, 2027. | Series: Scientific American investigates gems | Includes glossary and index.
Identifiers: ISBN 9781538314234 (pbk.) | ISBN 9781538314241 (library bound) | ISBN 9781538314258 (ebook)
Subjects: LCSH: Emeralds–Juvenile literature.
Classification: LCC QE394.E5 F625 2027 | DDC 553.86–dc23

Designer: Tanya Dellaccio Keeney
Editor: Therese Shea

Photo credits: Cover Yahor Mikhalenka/Shutterstock.com; p. 5 (top) KrimKate/Shutterstock.com; p. 5 (bottom) photo-world/Shutterstock.com; p. 7 (top) https://upload.wikimedia.org/wikipedia/commons/d/df/Béryl_var._émeraude_sur_gangue_(Muzo_Mine_Boyaca_-_Colombie)_2.jpg; p. 7 (bottom) VectorIcon2003/Shutterstock.com; p. 9 NPDstock/Shutterstock.com; pp. 11, 13 Jhampier Giron M/Shutterstock.com; p. 15 Lovember06/Shutterstock.com; p. 17 (bottom) Rachmo/Shutterstock.com; p. 17 (top) Angel Orjuela/Shutterstock.com; p. 19 STUDIO492/Shutterstock.com; p. 21 https://upload.wikimedia.org/wikipedia/commons/8/89/Gachala_Emerald_3526711557_849c4c7367.jpg; p. 21 (bottom) https://upload.wikimedia.org/wikipedia/commons/f/f3/Hooker_Emerald_Brooch.jpg.

Printed in China

CPSIA compliance information: Batch #QSSA27. For Further Information contact Rosen Publishing at 1-800-237-9932.

CONTENTS

Words in the glossary appear in **bold** type the first time they are used in the text.

AMAZING EMERALDS

Emeralds are green gemstones. Conditions within Earth must be exactly right for crystals of emerald to take shape.

Emeralds are one form of the **mineral** beryl. Pure beryl is clear and without color. It's made up of certain **elements**, including beryllium and aluminum. When the elements chromium or vanadium mix with beryl's elements, they may replace the aluminum. This gives beryl a green color. This can happen when superhot liquids move and mix elements beneath Earth's surface. It doesn't happen often. Emeralds are rare!

FUN FACT

THE OLDEST EMERALDS IN THE WORLD ARE NEARLY 3 BILLION YEARS OLD!

This is a raw emerald before it's cut into jewelry form. Colorless beryl, called goshenite, is shown above.

WHERE IN THE WORLD?

Ancient records tell us that emeralds were mined over 4,000 years ago in Egypt. Today, many say the finest emeralds come from the country of Colombia in South America. The Inca people discovered emerald **deposits** there long before Spanish explorers arrived.

The Spanish wanted to find more emeralds and dug many more tunnels. They **enslaved** the Inca to work for them. Hundreds of years later, emeralds are still found in some of these mines. The Muzo and Chivor mines are the most famous.

Emeralds are found in many places around the world. This map shows the biggest producers.

DESCRIBING EMERALDS

Emeralds usually form as six-sided crystals. They come in different shades of green. For example, Colombian emeralds tend to be deep green, while Zambian emeralds are often bluish-green. The element iron may affect an emerald's shade of green.

As the emerald crystals grow, other minerals can get trapped inside them. Bubbles of gas or liquid may also get trapped. These minerals and bubbles are called flaws or inclusions. Emeralds with many flaws are cloudy. The most valuable emeralds have few flaws and are clear. Most natural emeralds have flaws, however.

You can see the inclusions in these emeralds. Inclusions look like plants to some. They're sometimes called jardin, which means "garden" in French.

FUN FACT

PEOPLE BEGAN CREATING LAB-MADE EMERALDS IN THE 1960S. THEY'RE EASY TO SPOT BECAUSE THEY HAVE NO FLAWS!

TUNNELING INTO THE EARTH

To reach deposits of emeralds, miners may dig tunnels deep into the ground or into the sides of mountains. It's hard, unsafe work as tunnels can fall in. In the mine, emeralds can be hard to see, and it's easy to break them. Miners use hand tools to carefully dig through the rock and dirt.

Rock and dirt piles are taken to the surface in carts. Then they're carefully washed while the miners look for emeralds. Many carts of rock and dirt may have to be removed to find a few gem-**quality** stones.

Sometimes, **explosives** are used to blast into rock in emerald mines.

FUN FACT

SCIENTISTS CALLED **GEOLOGISTS** FIND WHERE TO LOOK FOR EMERALDS. THEY DRILL INTO THE GROUND WHERE THEY THINK CONDITIONS WERE RIGHT FOR EMERALDS TO FORM.

IN THE OPEN PIT

Most large companies look for emeralds through a method called open-pit mining. They make a series of steps, or terraces, as they dig down with large machines. This creates a large bowl-like shape in the ground.

Many miners often work on small pieces of land close to each other. Open-pit mining needs a lot of water. Miners dig out rock and dirt. Water helps them wash away the dirt to better see if the rock contains emeralds.

You can see the path of emerald formation in this mine in Columbia.

BECOMING JEWELRY

When an emerald is found with a rich color and few inclusions, it's cut and **polished** to make into jewelry. Even though emerald is hard, it can be split easily where it has inclusions.

Jewelers use a diamond saw and lots of water to cut the emerald crystal away from the rock it's found in. They go slowly so they don't harm the stone. The larger the finished emerald, the more valuable it will be. However, a smaller, grass-green emerald with fewer inclusions is more valuable than a large, pale stone with many inclusions.

Jewelers who cut and shape raw crystals are called lapidaries. They have lots of special tools.

FUN FACT

EMERALDS ARE GRADED. THE HIGHEST QUALITY GRADE IS AAAA.

AT THE WHEEL

Lapidaries don't hold the emerald in their fingers. They use a strong, sticky wax, called dop wax, to hold the raw emerald on a stick. The lapidary holds the emerald against a **grinding wheel** and makes flat sides called facets around the stone. Facets have to be evenly spaced on a finished gem.

Emeralds are usually cut into square or rectangle shapes. Jewelers think this shape best shows color and sparkle. One kind of cut with long **parallel** facets is called the emerald cut.

This lapidary holds the emerald on a grinding wheel using a dop stick.

FUN FACT

EMERALDS ARE ALSO USED IN SOME KINDS OF ELECTRONICS AND OTHER MACHINERY.

WHAT MAKES THEM VALUABLE?

An emerald's value has to do with its color and cut, but also its **clarity** and carats, or weight. The fewer inclusions an emerald has, the easier it is to see through it. This gives the emerald a high level of clarity.

A unit called the carat is used to weigh emeralds and other gems. Five carats are equal to 1 gram, or 35/1,000 ounce. A high-quality emerald is more valuable than a diamond of the same weight.

FUN FACT

IF THE FACETS ARE ALL EVENLY SPACED, AN EMERALD SPARKLES WITH A COLOR JEWELERS CALL "GREEN FIRE."

Flawless, deep-green emeralds are the most expensive gems in the world.

FAMOUS EMERALDS

Some high-quality emeralds have become famous throughout history. The Hooker Emerald Brooch was one of the most flawless emeralds. In the 1800s, it was worn by a sultan, or ruler, of the **Ottoman Empire** in his belt buckle. Actress Elizabeth Taylor had an emerald and diamond **brooch** that is the costliest emerald jewelry ever. It sold for $6.6 million in 2011.

The Bahia Emerald became the largest single crystal ever found in 2001. It weighs about 752 pounds (341 kg), or about 189,600 carats. What amazing emerald discovery is next?

GACHALÁ EMERALD

Mined in Colombia, the Gachalá Emerald, weighing 858 carats, is kept at the Smithsonian Institution museum in Washington, D.C.

HOOKER EMERALD BROOCH

GLOSSARY

brooch: A piece of jewelry that can be pinned to clothing.

clarity: In gems, free from flaws, also known as inclusions and blemishes, inside and outside.

deposit: An amount of a mineral in the ground that built up over time.

element: Matter, such as chromium, that is pure and has no other type of matter in it.

enslave: To force someone to work without pay under terrible conditions.

explosive: A kind of matter or device that produces a powerful blast.

geologist: Someone who studies the history of Earth, especially its formations and rocks.

grinding wheel: A wheel with a rough surface, used to shape or smooth a gem.

mineral: Matter in the ground that forms rocks.

Ottoman Empire: Relating to the Turkish political power begun in the late thirteenth century in western Asia. It ended in 1922.

parallel: Describing lines that are the same distance apart at all points.

polish: To make something smooth and shiny by rubbing it with a soft cloth.

quality: The standard or grade of something.

FOR MORE INFORMATION

Books

Amin, Anita Nahta. *Dig and Discover Crystals*. North Mankato, MN: Capstone Press, 2023.

DK Staff. *Rocks & Minerals*. New York, NY: DK Publishing, 2024.

Websites

Crystal
kids.britannica.com/kids/article/crystal/399416
Read more about the natural process of crystallization.

Emerald
gemkids.gia.edu/gem/emerald
Learn about emeralds and other gems on the GemKids site.

INDEX